Rock Brothers

Patricia Pike

ISBN: 978-0-473-48441-5

Other books by Patricia Pike

Muddled Memories
Underground rock stars
Bat and the glow worm
Air Whisperers of Nkandla
Final Waltz

The breeze wafted across the waters of Lake Taupo as Marama and her Nana walked along its banks.

Soft little fingers reached for her grandmother's hand searching for reassurance as Marama sobbed. "Nana. I'm a bad girl Nana." Tears streamed down her face as Nana listened to her sad tale.

"Ah, no sweetheart. You could never be bad. Why are you crying? Tell Nana and we can work it out together."

Nana lifted Marama onto the garden bench, placed to have a clear view of the mountains. Her granddaughter wriggled onto the seat and it gave the child time to think about what she would say. She whispered as tears continued to fall, "I don't want Mommy and Daddy to bring home another brother. I don't even like the one I have. Hemi can't talk yet and he doesn't like playing with dolls. I don't think I can love another baby. I must be bad because Mommy says loving is important." She looked up at her grandmother with her heart in her eyes.

PRIVATE

"Mmm, loving is not always easy Marama. And brothers can be really annoying sometimes. Hemi will grow up and be able to talk but I don't know if he will ever enjoy playing with dolls. Your Dad played with dolls. Yes, truly he did. They were all GI Joe dolls that had guns and boots and stuff and he treated them very differently to how you treat your dollies. But he loved his toys. Maybe this new brother will be different and enjoy a tea party with you and your girls? Or maybe he will only play with Hemi. But at the end of the day you will learn to love your brothers, sweetheart. I know this because you have a very big heart. A heart that could move mountains." Nana hugged the little girl to her side as they gazed out across Lake Taupo.

"Let me tell you about seven giant brothers who live in this area." Nana smiled as she looked into the distance. Nana didn't notice that Marama frowned in concern as she imagined giants roaming the land where she lived. The only giants the child had heard about were nasty and angry.

"Before men lived in Aotearoa, which we now call New Zealand, the giant mountain brothers all fell in love with one woman." Nana smiled as she noted her granddaughter's frown. "This is a story about a whole bunch of boys and how they dealt with their problems. They fought. Boys do that sometimes even if they like each other." Pulling a fern frond from the garden she tickled her granddaughter until Marama grinned and relaxed.

"Oh Nana, this story sounds awful. Can't we have another story about fairy princesses or mermaids? With ponies or soft teddy bears. Please Nana," Marama begged.

"You will enjoy this story Marama. It tells the story about our very own volcanoes." She pointed across the lake. "That is Tongariro and Ruapehu and there are other ones close by. We even get our electricity from a volcano. Before man walked these lands and no one in the whole wide world even knew that this country existed; the volcanoes were growing in strength."

Nana paused for a moment before saying, "Each brother fought for the love of the beautiful fern covered lady-mountain. She was soft and kind and everyone wanted to be the one she would choose to be with forever. Her name is Pihanga."

"Oh Nana, I have seen Pihanga and she is beautiful. She is full of ferns and forests. Okay Nana, you can tell me your story. I like Pihanga and I always knew she was a girl mountain." Marama made herself comfortable as she settled down to listen to the legend.

"Each brother fought the other. They threw fire and rocks and smoke at each other. They grumbled and roared and shook the earth from one end of the country to the other. But one mountain giant was stronger than the rest. He was Tongariro." Nana pointed to the low profile of the active volcano still visible at the South end of Lake Taupo. "Ngaruhoe and Ruapehu took Tongariro's side in the fights and stood back. Tauhara moved to the North and even Putauaki was defeated. In the end Pihanga promised to be the wife of Tongariro. And he was very pleased for he truly loved Pihanga. Some of the brothers moved away because they did not want to see the lovely Pihanga marry their strong sibling."

Nana sighed and said, "This happened in the year 186 AD, which is a long, long time ago. Long before even Nana was born. Or my Nana or even her Nana." Nana sat back and smiled at the memories that came to mind of sitting on the lap of her own Nana all those years ago when she was as young as Marama was now. "And all was well for a while. Pihanga gave birth to many children to Tongariro and he was happy. Tongariro would go off hunting and Pihanga would stay behind to care for their family."

Marama said, "Mom stays home while Dad goes to work. Is that the same Nana?"

"Well, not really because when your Dad is at home, he spends lots of time with your Mom and you children. Tongariro didn't and Pihanga felt sad and forgotten. I imagine that Pihanga would have loved to have a husband like your Dad. Pihanga started to feel neglected and thought that Tongariro didn't care about her, so she fell in love with another mountain giant. Taranaki."

Marama jumped up and yelled, "We have been to Taranaki, Nana but it isn't close to Taupo. That mountain is almost in the sea; far away."

"Yes, Marama. You're so clever. Taranaki is far away and that is part of the story," Nana answered with a smile. "Pihanga had another baby and this baby didn't look anything like the others she had with Tongariro. He started to think she was cheating on him and had another giant in her life." Nana frowned slightly at the thought of the turmoil in days of ancient history.

"Tongariro decided to pretend to go to sleep so that he could spy on his wife. He closed his eyes and Pihanga thought that he really was asleep." Nana closed her eyes and pretended to snore. "As soon as she thought Tongariro was asleep Pihanga embraced Taranaki who was the Daddy to her littlest volcano baby."

Marama giggled and then said, "Oh no Nana. Did they get divorced? My friend Sarah's parents got divorced and she cried and cried and had to move away from her big house. She lives in a small house now and is very sad because she hardly ever sees her Daddy." Marama looked shocked at this twist in the story.

"No Marama. It was worse than that. Tongariro was so angry that he fought with Taranaki. He threw stones and ash into the air and finally Taranaki chopped off the head of Tongariro and it landed in the middle of Lake Taupo. We call it Moututaiko Island. But Tongariro was really the stronger of the two mountains and even without his head, he won the fight." Nana took a deep breath as she considered the battle that had been fought.

"Ah, Nana I don't believe that his head landed in the lake. That sounds weird. Was the battle that bad?" Marama asked.

"Oh yes, sweetheart. The ash flew up into the sky for fifty kilometers. It went to where the sky meets the stars, in outer space. It was huge. The ash and heat flew from Tongariro at nine hundred kilometers per hour. Even a very fast car could not drive quick enough to escape. Not even a cheetah or a hawk could escape that speed. The ash and pumice fell and fell until everything was covered, even as far as Auckland and Wellington. This house would have been blasted into nothing and then covered as high as ten meters. Do you see that Pine tree? That is ten meters tall at least and it would have been buried in hot burning ash. Not nice."

10

Nana shivered as she imagined that far distant day. Marama shook her head as she tried to envision the forces of nature at work. "And Pihanga, Nana, was she alright?" Her voice trembled as she asked, "No sweetheart. She wasn't okay. Her ferns and forest were burned to a crisp. Taranaki grew sad to see the beautiful Pihanga so hurt. He stopped fighting to save her. He loved her very much but he knew he could not defeat Tongariro." Nana looked sadly at the horizon.

Marama put her hand on her Grandmother's arm and said, "Tell me more Nana. Please."

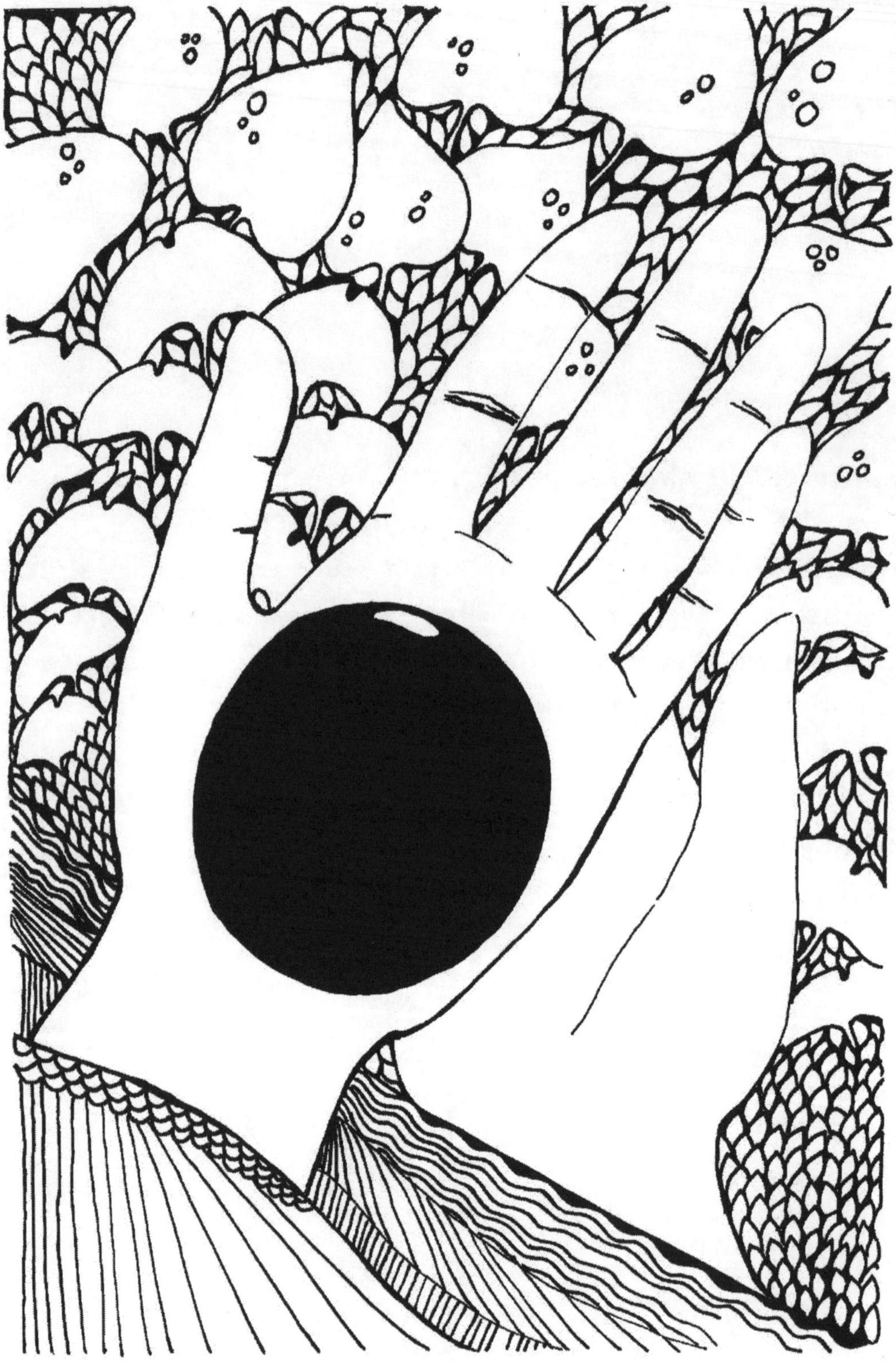

"Right," Nana said as she wiped a tear from her cheek. "Ruapehu told Taranaki to leave and to continue on until the end of the day. Where he stopped would be his resting place. Ruapehu gave him a magic stone, called Toka-a-Rauhotu to lead and guide him. And Taranaki left taking his baby with him. The baby's name is Panitahi and she is with him still. He dived under the earth as he escaped, gouging out the Whanganui river bed as he rushed off. He was a warrior and brave as any other mountain. But he knew he could not win and decided it was better to retreat."

Nana hugged Marama to her side. "He went to the coast where he met another lady mountain, Pouakai. They have many children together. I like to think he found happiness after all," Nana said, a grin touching her lips. "Toka-a-Rauhotu has the power to release Taranaki from his frozen state and maybe one day he will. That would be fun to see. A volcano walking across the land sounds very exciting. I wonder if he will return to Pihanga and if Tongariro will be angry again?"

Nana smiled. "Tongariro wraps his lady love in clouds to protect her. He doesn't go off hunting anymore and she no longer feels neglected. But he sometimes remembers her betrayal and growls and rumbles when he thinks she is sending signals to her lover. Taranaki hasn't forgotten Pihanga and he sends smoke through the sky to remind her that they once shared a great love." Nana shook her head at the idea of sending messages to a lost love. "Tongariro cries tears of sadness and both the Waikato River and the Whanganui River start with him."

"Oh, Nana that is a lot of crying. Poor Mountain to have so many tears that it fills two rivers. Have you seen Huka Falls, Nana? And that is only one part of the tears he has shed. That makes me so sad. But it is nice that he still loves Pihanga," Marama affirmed.

"True. Love can conquer all. And yes, Marama, I have seen Huka Falls and how it falls over a waterfall into the Waikato River. I love visiting the Falls. It's a beautiful place to think and enjoy nature. Maybe we should go for a picnic one day when you have to escape from your brothers." Nana laughed softly.

Marama shook her head in excited agreement.

"What this story tells us is that love is not always easy. But if we just try our very best, things will work out and we can all live happily ever after." Nana patted Marama's hand. "Yes Marama, you can love your brothers. And let us hope that they never fight like the volcanoes did two thousand years ago." Nana hugged Marama as they watched a wisp of smoke rise from the top of Tongariro.

"Look Nana, he is talking to us. And he thinks we should have a picnic right now." Marama cajoled her Nana. "Some of those nice biscuits Nana, and maybe a cold drink?"

"Oh okay. Let's wake up your brother, Hemi, and have a little picnic while we wait for your parents to bring the baby home. I know you will love your new baby brother, sweetheart. Just like we all fell in love with you when your parents brought you home a few years ago." Nana kissed the top of the child's head as they strolled back into the house. "You couldn't speak then and just look at you now. Chatting away ten to the dozen."

"The story goes on Marama," Nana said as she put biscuits on a plate.

“Many, many years later a man called Ngatoro-i-rangi came here on a voyage of discovery. He climbed up Ruapehu with a servant. You know how cold it can get in the snow and Ngatoro-i-rangi had come from a warm island home and was not prepared. The clothes they wore were made from grass and palm leaves. Not very warm.” Marama shuddered and leaned into her side as she continued to explain. “No shoes. Can you imagine walking on the snow with no shoes. The servant got colder and colder. They got so very cold that the servant he got sick.”

Ngatoro-i-rangi was a Maori priest or tohunga. That meant he had some power.”

“You mean like magic?”

“I suppose you could call it that. So he called out to his sisters far away across the ocean to send him fire, to keep them warm. But before the fire could reach them the servant fell asleep and never woke up.”

“Now this is a terrible story Why can’t he be alright?”

“Sometimes terrible things happen to nice people.”

Marama said, "I went to the snow and I had to wear lots of clothes Nana. Didn't Ngatoro—um um, I can't say his name properly, have a puffer jacket or anything Nana? Or boots or even a nice jersey? Or even plastic bags around his feet to keep them dry? I do that. I put plastic bags around my socks and then put my boots on and my feet are always dry at the end of the day."

"No, sweetheart. And he shivered and shivered with cold. His sisters heard his call and sent fire to keep him warm. That fire is seen in a volcano off the coast of Tauranga called White Island where it rose out of the sea to find its way. Then it rose out of the ground at Rotorua and Tarawera and finally the fire reached Ngatoro-i-rangi and warmed him up." Nana poured the juice into frosty glasses as she spoke.

“Each

tribe has a different story about the volcanoes. But I like the one I told you. I know that love can move great mountains and that peace can be found amidst turmoil if you really try to discover it," she concluded.

Hemi wandered into the kitchen rubbing the sleep out of his eyes and they lit up as he saw the picnic being laid out for them. He gave his Nana a big hug and then sat next to his sister. He looked up at her and lisped, "Love you too, Arama." As he stuffed a biscuit into his mouth.

ABOUT THE AUTHOR

Patricia Pike a former South African now living in rural New Zealand. She lives with her husband and her two cats. Her grown children live all around the world and also close by. Family is very important to her.

She has been interested in art all her life. Story telling has just evolved over time.

Living near the Waitomo caves has given her an interest in the local legends and stories.

THANK YOU

Made in United States
Troutdale, OR
11/03/2026

56845264R00020